OUT GRAVITY

FLOATING THROUGH FATE

GIRIJA ARUMUGAM

About the Book:

In a world that has always been grounded by the laws of nature, what happens when those very laws are defied? "Out of Gravity: Floating Through Fate" is a gripping exploration of a planet unshackled from its gravitational bonds. Dive deep into tales of suspended cities, airborne cultures, and the daily struggles and joys of a populace floating amidst the clouds. This novel is a lyrical blend of science, romance, and adventure, painting a vivid tapestry of a world where boundaries are redefined and the human spirit soars.

About the Author:

Girija Arumugam is an acclaimed author known for her ability to weave intricate stories that straddle the realms of reality and fantasy. With a background in astrophysics and a heart rooted in storytelling, Arumugam crafts narratives that not only captivate but also make readers ponder the deeper mysteries of existence. "Out of Gravity" is her latest offering, a testament to her unparalleled imagination and her profound understanding of the human experience.

Copyright

Table Of Contents

Table Of Contents

Table Of
Contents

Table Of
Contents

Book Introduction:

Imagine a world where your feet never touch the ground, where buildings, trees, animals, and even droplets of water float freely in the atmosphere. This is not the setting of a sci-fi movie nor an imaginative dream. This is Earth, and it has lost its gravitational pull.

Out of Gravity: Floating Through Fate takes us on a journey through this altered realm, exploring the immediate and long-term consequences of a world without gravity. The once familiar planet has transformed into a sphere of chaos, as everything that was once grounded is now suspended in the air. The serenity of blue skies has become the tumultuous ocean of floating obstacles.

Book Introduction:

How would we adapt when the anchor that has held us for millennia suddenly disappears? The challenges are endless: from the basic act of consuming water to the complexity of maintaining relationships, businesses, and social structures. How does one embrace love when a simple hug is no longer grounded? How do economies function when the very basis of trade is disrupted?

Every facet of life as we know it has been radically altered, pushing humanity's resilience, creativity, and spirit to its limits. But as we will discover, it also paves the way for new beginnings, innovations, and a redefinition of life's priorities.

The Moment Gravity Disappeared

The sun rose like any other day, casting a golden hue over the city's skyline. Children were preparing for school, adults were gearing up for work, and the world was bustling in its routine manner. But at precisely 8:32 AM, everything changed.

Sarah, a young journalist, was sipping her morning coffee when her cup, along with everything else in her apartment, began to levitate. Confused and alarmed, she tried to grab her phone but found herself floating upwards, colliding with the ceiling.

Across the city, vehicles started drifting off the roads, causing panic among drivers. Birds, used to soaring freely, struggled to navigate the sudden change, often colliding with floating objects or people. Fish from ponds and rivers were suspended in air pockets, gasping for oxygen.

News broadcasts, those that could still transmit, painted a picture of global confusion. No corner of the Earth was untouched. Videos streamed in from around the world showing iconic landmarks like the Eiffel Tower or the Pyramids of Egypt surrounded by a whirlwind of floating debris and awestruck people.

Panic ensued as everyone scrambled to find a way to anchor themselves. Trees became temporary shelters, with people clinging to them. Those inside buildings tried to tie themselves down with anything they could find: belts, ropes, and cords.

But as the initial shock began to wane, the reality of the situation settled in. The world had changed forever, and the age-old belief that 'what goes up must come down' was no longer valid.

Adapting to the Air: The First Few Hours

The immediate hours following the loss of gravity were crucial. Without any warning or preparation, humanity had to quickly figure out basic survival.

Breathing became the first challenge. With a massive amount of dust and particles lifted into the air, many found it difficult to breathe. Makeshift masks from cloth and clothing were used to filter out the larger particles.

Communication was vital. Families and friends sought each other out, calling out names and creating human chains to stay together. The sound of voices, music, and even loud noises became beacon points, guiding people towards safe zones.

As the day progressed, some entrepreneurs saw an opportunity amidst the chaos. Floating markets emerged, where items such as bottled water, masks, and ropes were exchanged or sold. Despite the absence of a solid ground, human ingenuity began to shine.

The Floating Cities: Sky Architecture

With the understanding that this might be the new normal, there was an urgent need for shelter. Traditional houses and buildings were no longer feasible. The first floating settlements were rudimentary, often just a group of objects tied together, forming a makeshift platform.

However, as days turned into weeks, these floating platforms evolved. Architects and engineers, working with available materials, began designing floating homes that utilized aerodynamics and balance to create stable living spaces.

These 'Sky Homes' became the new architectural marvels. Using lightweight materials and creative designs, they provided shelter, safety, and a semblance of normalcy.

The Contest of Nutrition: Eating and Drinking Flooded

In a gravity-less world, the simple act of consuming food and water became an adventure in itself. What was once a straightforward task of lifting a spoon to one's mouth or pouring water into a cup became a dance of dexterity and innovation.

The Liquid Dilemma:

Water, an essential element of survival, posed the first challenge. Without gravity, it didn't pour; it floated. Drinking from a cup was no longer possible as droplets of water would simply drift away. The sight of floating spheres of water became common, with people chasing after them, trying to catch and swallow these bubbles before they escaped.

Innovation came quickly. Companies began producing sealed drinking pouches with straw-like attachments. By squeezing the pouch, water would be pushed into the mouth, allowing for controlled drinking. This became the standard method of hydration.

Solid Foods and the Floating Feast:

Eating solid foods required equal parts creativity and patience. With no plates or bowls to hold food, and no way to keep items from floating away, people began using skewers or sticks to eat. Kebabs became an unexpected staple, as they were easy to hold and prevented food from drifting away.

Sandwiches and wraps, foods that were self-contained, saw a rise in popularity. Floating markets sold various wraps filled with meats, vegetables, and sauces, providing a nutritious and manageable meal option.

The Nutritional Balance:

With traditional farming disrupted and livestock floating aimlessly, there was an immediate concern about maintaining a balanced diet. Hydroponic farms, previously a niche industry, became essential. These farms, which grew plants without soil, were adapted to function in a zero-gravity environment. Floating pods of nutrient-rich water nourished plant roots, allowing for the cultivation of vegetables and fruits in mid-air.

Livestock farming became more challenging. Some animals, like chickens, could not survive long in these conditions. As a result, people shifted to consuming more fish, which were kept in sealed, water-filled pods and proved easier to manage.

Tethered Homes: The New Way to Stay Grounded:

In a world where everything floated freely, the concept of 'home' underwent a drastic transformation. As Earth's population adjusted to the new state of weightlessness, the basic human need for shelter and a sense of belonging became paramount. Thus began the era of the tethered homes.

The Birth of Tethered Communities.

In the initial chaos post the loss of gravity, many found themselves adrift with their entire homes and belongings. The idea of tethering homes was born from the urgent necessity to prevent properties from floating away into the vast expanse of the sky.

The first tethered homes were makeshift solutions. People used ropes, chains, and even repurposed vehicle seatbelts to anchor their floating abodes to large, sturdy structures like skyscrapers, bridges, and mountains. Communities began forming around these anchor points. Over time, these random congregations of floating homes took on more structured forms, reminiscent of neighborhoods and suburbs, but in the sky.

Architectural Marvels in the Sky:

With the realization that this could be a prolonged or even permanent scenario, architects and engineers collaborated to design homes specifically for this new world. These homes were lighter, built with materials that could withstand varied weather conditions, and designed with tethering points that allowed them to be anchored securely.

The tethered homes were interconnected with flexible walkways, allowing residents to move from one home to another without the fear of drifting away. These homes often had multiple levels, taking advantage of the vertical space, with windows on all sides providing panoramic views of the floating world.

Economies of the Sky:

The tethered communities spurred new economic opportunities. Entrepreneurs set up floating markets adjacent to these homes, selling essentials from food to clothing. Delivery services adapted, with drones becoming the preferred method to transport goods from one tethered community to another.

Local governance structures formed within these communities, with leaders being elected to ensure the safety and well-being of the residents. There were community gatherings, floating parks for children, and even schools that adapted to this aerial lifestyle.

The Social Fabric of Tethered Living:

Living so closely with neighbors, both horizontally and vertically, reshaped social dynamics. Privacy became a luxury, and communal living became the norm. Shared gardens, entertainment hubs, and recreational areas became common in tethered communities.

Despite the close quarters, or perhaps because of it, a deep sense of camaraderie developed. Festivals, celebrations, and events took on a communal spirit, with everyone contributing and participating. Love stories blossomed between adjacent homes, and tales of sky-born adventures became popular bedtime stories for children.

Navigating Challenges:

Life in tethered homes was not without its challenges. Weather became a significant concern. Strong winds could strain the tethers, and storms became doubly hazardous in the sky. Engineers constantly sought ways to enhance the security and resilience of tethering mechanisms.

Space became a premium. As more and more people sought the security of tethered communities, disputes over tethering spots and space arose. Negotiations, bartering, and sometimes conflicts occurred over prime tethering real estate

The tethered homes represented more than just a practical solution to the gravity-less state of the world. They epitomized humanity's resilience, adaptability, and unyielding spirit. In floating homes tethered to the remnants of the old world, humanity found a way to rebuild, reimagine, and rekindle the sense of home and community.

The Rise of Airborne Cultures

A New Language in the Air:

The weightlessness introduced a unique vernacular into everyday speech. Words like "drift-dancing" described the weightless movement of people, and "sky-walking" became a term for navigating through floating obstacles. "Cloud-surfing" referred to the act of moving along dense patches of suspended water vapor, and "tether-talks" became popular social events where people would tether themselves to a spot and chat.

Fashion Beyond the Ground:

Fashion underwent a renaissance. Designers saw a canvas like never before. Outfits were tailored to allow ease of movement in weightlessness, often integrated with pockets of air or small propulsion devices to assist in navigation.

Shoes transformed from being mere foot protection to tools, featuring small magnets or hooks. Accessories were not just about aesthetics; they had safety functionalities like wrist-tethered bags and floating hats equipped with mini fans to steer direction.

Artistic Expressions:

Art became an ever-evolving medium. Painters embraced the challenge of creating in a weightless world, where paint droplets could float, resulting in 3D masterpieces. Sculptors started using lighter materials, creating floating installations that people could move through, interact with, and experience from all angles.

In literature, a new genre of 'Sky Fiction' emerged, with stories weaving tales of love, adventure, and mystery set against the backdrop of the floating world. Poets romanticized the weightlessness, penning verses that resonated with the collective experiences of people.

Music and Dance: The Weightless Ballet:

Musicians found that sound behaved differently without gravity. Instruments were modified and new ones were invented. The "Aero-Harp," a stringed instrument that used air currents to produce sound, became a favorite. Music genres like "Drift Beat" and "Cloud Core" emerged, encapsulating the essence of the floating world.

Dance, previously bound by the laws of gravity, transformed into an aerial performance. Dancers, tethered to floating platforms, would perform routines that seemed to defy logic. Dance troupes like "The Airbound Ballet" and "Zero-G Groovers" gained global fame, with their performances becoming highly sought-after events.

The Floating Festivals:

In this new world, celebrations too found their place in the sky. The Floating Lantern Festival became an annual event where people released bioluminescent organisms into the air, creating a spectacle of dancing lights. The Zero-Gravity Music Fest drew artists from around the world, each bringing their unique airborne tunes.

The rise of airborne cultures was not just about adapting to a new environment; it was a testament to human resilience and creativity. It highlighted the indomitable spirit of humanity to find beauty, expression, and connection, even when the very foundations of their world had shifted.

Love in Zero Gravity: Relationships Reimagined :

The Gravity of First Impressions:

In a world without gravity, first encounters took on a whimsical charm. It wasn't uncommon for love stories to begin with tales of accidental collisions amidst floating streets or shared laughter while chasing a runaway object. These unplanned meetings in mid-air often blossomed into beautiful relationships, each anchored not by gravity but by shared experiences and emotions.

Floating Dates and Skybound Adventures:

Traditional date night activities were reimagined for the skies. Restaurants offered tethered dining experiences where couples could enjoy meals while floating above scenic landscapes. "Cloud cafes" emerged, where partners could sip on their favorite beverages amidst soft, floating clouds. Cinema nights turned into immersive experiences, with couples floating within holographic movie scenes.

Nature dates reached new heights — literally. Floating gardens became popular spots for romantic rendezvous, with couples drifting among blossoms and bioluminescent plants. Watching meteor showers took on a literal sense as couples would tether themselves on high platforms to get closer to the stars.

The Dance of Intimacy:

Physical connections became an art. Holding hands wasn't just a sign of affection but also a means of staying together in the floating world. Hugs became a safety embrace, ensuring neither drifted away. Kissing, a delicate balance of approaching and retreating, became a dance in itself. Couples developed their unique rituals and gestures to express love and intimacy, blending affection with the practicalities of a zero-gravity world.

Challenges of the Heart:

Like any relationship, couples faced challenges. The vastness of the floating world meant that distance took on a new meaning. A simple task like visiting a partner's home could become a daring adventure, navigating through floating obstacles and unpredictable air currents.

Safety concerns also added a layer of complexity. Couples often found themselves in situations where they had to act quickly to ensure each other's safety, be it tethering a drifting partner or shielding them from floating debris.

Deepened Bonds:

Despite these challenges, or perhaps because of them, relationships in the zero-gravity world deepened. Couples relied heavily on trust, understanding, and communication. The shared experiences of navigating the weightless world, the daily adventures, and the moments of vulnerability brought them closer than ever.

Engagement rituals evolved. Instead of rings, couples exchanged personalized tether bracelets, symbolizing their commitment to staying connected and grounded to each other, irrespective of the world around them.

Love, in the gravity-less world, transcended the physical realm. It became a force that kept people grounded, provided comfort in the chaos, and instilled hope in the heart. While gravity had disappeared, the pull of love remained as strong as ever, reminding everyone that true connections were not bound by the laws of nature.

The Economics of Floating: Business and Trade Upended:

Brick-and-Mortar: A Relic of the Past:

As Earth became a vast expanse of floating entities, the very foundation of traditional businesses crumbled. Storefronts that once proudly stood on street corners now hovered aimlessly, their merchandise at the mercy of the new environment. Businesses had two choices: adapt or perish. Many chose the former, transforming their operations to cater to a weightless world.

Drone Deliveries: The New Norm:

In a realm where roads became obsolete, the sky became the primary route for transportation. Drones, once a budding industry, suddenly became the backbone of global commerce. Enhanced with powerful propulsion systems and advanced navigation technology, they darted between floating homes, tethered communities, and drifting marketplaces, delivering everything from food to fashion.

Entrepreneurs saw potential and launched drone-based startups, offering services like 'Quick Drift' for instant deliveries and 'SkySave' for emergency supplies

Financial Fluctuations in the Floating World:

While stock markets had already transitioned to the virtual realm, their dynamics shifted with the new airborne reality. Companies producing anti-gravity equipment, tethering devices, and weightless-friendly products became the blue-chip stocks.

Traditional real estate companies faced declines, while businesses that specialized in floating architecture or tethered communities saw exponential growth.

Cryptocurrencies gained even more traction, given their digital nature and the practical difficulty of using physical currency in a world where coins and notes could simply float away.

Tourism Reimagined: Floating Wonders:

Tourism, a sector deeply rooted in Earth's landscapes and cultures, underwent a metamorphosis. Grounded landmarks like the Eiffel Tower or the Great Wall, though still significant, were now part of a larger, mesmerizing aerial panorama.

New attractions emerged: guided tours through floating forests, skydiving without the dive (a free-float experience in open skies), and visits to renowned floating communities.

Travel agencies offered packages like "Cloud Cruise" or "Starlit Soiree", giving travelers an experience of the floating world's wonders.

Emergence of Floating Marketplaces:

Without stable ground, a new kind of marketplace emerged. Floating bazaars, tethered at various altitudes, became hubs of commerce. Sellers showcased their wares in tethered stalls, and buyers navigated these markets with tethered baskets, filling them with goods as they floated from one stall to the next.

Challenges and Opportunities:

This airborne economic system wasn't without challenges. Regulatory frameworks had to be reestablished to ensure safe and ethical business practices. Disputes over airspace for drone deliveries, concerns over floating market stability, and the need for new business insurance policies covering 'drifting damages' were just some of the complexities the business world had to navigate.

Yet, with challenges came opportunities. Innovators thrived, introducing products and services tailor-made for the new world's unique needs. From floating farms producing fresh produce to entertainment companies offering zero-gravity sports and games, the business landscape was alive with potential.

In a world upended, the tenacity of human enterprise shone through. The economics of floating wasn't just about profit and trade; it was a testament to humanity's indefatigable spirit to adapt, innovate, and thrive.

Education in Elevation: Teaching the Next Generation

The School of the Skies:

As homes, businesses, and communities floated, so did schools. Buildings that once stood as educational pillars on the ground now became majestic hubs of learning in the sky. These institutions, aptly renamed "Sky Schools," were interconnected through flexible walkways, allowing students and faculty to traverse between different segments of the school.

Panoramic Classrooms: A Window to the World:

The traditional four-walled classroom was a concept of the past. Instead, classrooms boasted transparent walls, providing students with unparalleled views of the floating world. This ever-changing backdrop made each lesson unique. Geography classes utilized the views to point out real-time floating landforms, while literature classes took inspiration from the drifting clouds for poetry and prose.

Adapting the Curriculum:

Education had to be relevant to the challenges and wonders of the new world. Science classes focused on understanding the phenomena of the weightless Earth, probing the theories and possibilities of the sudden change. History lessons emphasized the grounded past, ensuring that the next generation remembered the world as it once was.

Physical education saw a transformation, with traditional sports being modified for the air. "Aero-ball," a mix of basketball and aerobatics, became a popular sport. Swimming lessons were replaced by floating navigation courses.

Essential Life Skills: Safety and Navigation:

Safety in this new environment was paramount. Schools introduced mandatory courses on tethering techniques, floating safety protocols, and emergency navigational tactics.

Children practiced drills, learning how to anchor themselves in sudden storms or how to use mini-propulsion devices to navigate their way home.

Innovation Labs: The Heart of Sky Schools:

Recognizing that the children, who adapted to this new world faster than adults, often had unique solutions to everyday challenges, schools incorporated Innovation Labs. Here, students could prototype their ideas, be it a new design for a floating backpack, a tethered shoe, or even communication devices optimized for the aerial world

These labs became incubation centers for the next generation of inventors and thinkers. Regular "Sky-Talks" were organized, where students presented their inventions to their peers, and the most promising ones were integrated into the community.

The Role of Virtual Learning:

Given the challenges of navigating the floating world, virtual learning saw an upsurge. Digital classrooms, augmented reality lessons, and holographic tutors became common. These platforms allowed students from different floating communities to interact, share experiences, and collaborate on projects, bridging the distances created by the vast open skies.

Education in elevation was not just about academics; it was about preparing the next generation for a world full of uncertainties and wonders. The floating schools instilled a sense of curiosity, resilience, and adaptability in students, ensuring that no matter where the winds of change took them, they would be ready to soar.

Wildlife Wonders: How Animals Adapted

Birds: Masters of the New Sky:

The birds, already rulers of the air, found themselves adapting to a world with different dynamics. No longer needing to expend energy in continuous flapping, many species evolved to master more intricate aerial maneuvers. Songbirds, for instance, began to use their floating environments to create harmonious calls that resonated differently, creating symphonies that echoed in the sky. Raptors like eagles and hawks developed hunting techniques to swoop down on floating prey with even more precision.

Marine Life: Navigators of Floating Oceans:

The seas rose and converged into magnificent floating water bubbles, transforming the aquatic life within. Fishes started to exhibit more three-dimensional movement, with species previously limited to the ocean floor now freely floating in the center of these water orbs. Whales and dolphins developed new communication patterns, their songs traveling through the interconnected water bubbles in a mesmerizing melody.

Coral reefs, now partially exposed to the open air, underwent a transformation. The symbiotic algae living in corals began to photosynthesize differently, resulting in reefs of brilliant, previously unseen colors.

Mammals: Agility in the Air:

Land mammals faced a unique challenge. Animals like squirrels, monkeys, and even larger species like leopards showcased unprecedented aerial agility. They hopped, glided, and even somersaulted from one floating land fragment to another. Elephants, with their massive size, became stabilizing anchors for larger floating islands, their trunks becoming adept at grabbing onto floating vegetation for food.

Domesticated animals, especially cats and dogs, became the companions humans needed in navigating the floating world. Their instincts and heightened senses often alerted humans to upcoming dangers or drifting resources.

Insects: The Tiny Titans of Adaptation:

The microcosm of insects displayed some of the most rapid adaptations. Bees built floating hives, using the free-floating flowers as their new nectar sources. Ant colonies constructed nests on drifting leaves, creating intricate networks that connected different floating plants. Butterflies, with their lightweight bodies, flitted between floating gardens with ease, their wings shimmering like stained glass in the ever-present sunlight.

Reptiles and Amphibians: A Shift in Habitats:

Cold-blooded animals like reptiles and amphibians found solace in the floating marshlands. Frogs and toads utilized floating water bubbles as breeding grounds, their tadpoles adapting to navigate in these enclosed aquatic environments. Snakes became adept at slithering between floating debris, and some even evolved to develop fringed scales, enabling them to glide short distances.

Symbiotic Relationships: A New Dance of Dependence:

In this floating world, many animals formed new symbiotic relationships. Birds often partnered with larger mammals, perching on them to rest, while in return, they would alert the mammals to any impending threats. Fishes cleaned the exposed parts of floating whales, ensuring they remained free from parasites.

The Earth's sudden transformation into a weightless realm posed countless challenges to its inhabitants. Yet, the resilience and adaptability of wildlife painted a picture of hope and wonder. Animals, in their unique ways, showcased that life, no matter the circumstances, always finds a way to not just survive but thrive. Their stories of adaptation became lessons in perseverance, innovation, and evolution, reminding all of Earth's interconnectedness and the boundless spirit of nature.

Celestial Celebrations: New Traditions Born:

The Festival of Floating Lanterns:

In homage to the drifting world, the Festival of Floating Lanterns emerged as a global favorite. Families would craft lanterns using biodegradable materials and infuse them with gentle glows using bio-luminescent organisms. Released into the night sky, these lanterns created rivers of light, symbolizing hope, unity, and the collective journey of humankind amidst the weightlessness.

Sky Waltz: The Dance Festival:

A celebration of movement in the weightless world, the Sky Waltz became an annual event. Participants, tethered to floating platforms, would engage in intricate aerial dances, moving to music that echoed through the skies. Dances told stories, from tales of love and separation to sagas of resilience and adaptation in the new world.

Harvest of the Air:

With traditional agriculture taking to the skies in the form of floating farms, the Harvest of the Air festival marked the season's yield. Communities would come together, sharing produce, and celebrating with feasts held on vast floating platforms. Traditional dishes were reimagined with the unique produce of the floating farms, leading to a culinary renaissance.

The Echoes: A Musical Tradition:

Music, an essential part of human expression, took on a new dimension. The Echoes became a celebration of sound in the open skies. Musicians would play on floating stages, their melodies resonating differently in the vast expanses. The festival would culminate in a global chorus, with participants from around the world adding their voices to a harmonious echo that circled the planet.

Drift Day: Remembering the Change:

Marking the day the world began to float, Drift Day became a solemn occasion of remembrance and reflection. People would tether together, forming vast human chains, symbolizing unity and shared strength. Candlelit vigils, poetic readings, and communal storytelling sessions became pillars of this day, ensuring that while the world moved forward, it never forgot its past.

Starbound Festival:

With the skies playing such a pivotal role in daily life, stargazing became more than a hobby; it turned into a celebration. The Starbound Festival, held during the clearest of nights, had enthusiasts and families lying on floating mats, observing constellations, and making wishes upon shooting stars.

Crafts in the Clouds:

Creativity found its way to the skies in the Crafts in the Clouds festival. Artists, designers, and craft enthusiasts would display their creations, from floating sculptures to intricate airborne tapestries. This celebration of art also became a marketplace, allowing artisans to trade and sell their weightless wonders.

As the world adapted to its new reality, so did its festivities. The celestial celebrations were not just about merriment but also about resilience, unity, and the human spirit's undying ability to find joy even in the most unfamiliar terrains. These new traditions served as anchors, grounding humanity in emotion, community, and shared memories amidst the ever-drifting world

The Search for Solutions: Scientists at Work:

The Global Assembly of Minds:

Shortly after the world began to float, a global consortium of scientists, engineers, and researchers was formed. Dubbed "The Aerial Alliance," this group's primary mission was to understand the cause behind Earth's sudden weightlessness and seek solutions to stabilize life in this new reality.

Research Labs in the Sky:

Traditional ground-based labs were replaced by state-of-the-art floating research stations. Equipped with advanced technology and tools, these labs became hubs of innovation, where experiments were conducted in real-time, taking advantage of the zero-gravity environment.

Probing the Phenomenon:

One of the Alliance's primary tasks was to decipher the cause of the global floatation. Hypotheses ranged from altered magnetic fields to cosmic interference. Deep-space telescopes were deployed, studying celestial bodies for clues, while ground samples were examined for any anomalous activity. Massive supercomputers, tethered together in a vast floating data center, crunched numbers and simulated theories, hunting for answers.

Engineering Life in Elevation:

Parallel to the investigations were engineering teams dedicated to making life sustainable. They developed anti-drift suits, allowing people to walk with a semblance of gravity. Agricultural pods, designed to cultivate crops in the floating world, were rolled out. Advanced tethering systems were created, ensuring stability of floating habitats.

Medical Marvels: Health in the High Skies:

Understanding the impact of prolonged weightlessness on human physiology was crucial. Medical researchers dove into studies, exploring everything from bone density changes to circulatory system adaptations. New medications and therapies were developed to counteract any negative health implications, ensuring that humanity could thrive in its new environment.

A World of Collaboration:

Perhaps one of the most significant changes in the scientific community was the heightened level of collaboration. With a global crisis at hand, barriers of competition and secrecy broke down. Data was shared freely, and multinational teams worked together, pooling resources and expertise.

The Youth Brigade: Fresh Perspectives:

Recognizing the importance of diverse perspectives, the Aerial Alliance initiated the Youth Brigade program. Young prodigies, thinkers, and innovators were enlisted to contribute fresh ideas. Many breakthroughs, from innovative floating transport systems to new energy harnessing techniques, were credited to these young minds.

Communication Channels: Keeping the World Informed:

With the world's population eagerly awaiting solutions, the Alliance ensured transparent and regular communication. Regular broadcasts, aptly named "Horizon Updates," were transmitted globally, informing people of the latest findings, safety guidelines, and innovations. These updates became a beacon of hope, assuring the populace that the brightest minds were tirelessly working for a better tomorrow.

The Search for Solutions was more than just a quest for answers. It was a testament to human resilience, adaptability, and the indomitable spirit of curiosity. In facing one of the most perplexing challenges in history, the scientific community rallied together, shining as beacons of hope and progress in an uncharted world.

The Emotional Strain: Mental Health Above Ground:

Floating Realities: The Emotional Whirlwind:

When the ground beneath vanished, it took with it a sense of security that humanity had always clung to. This new floating existence, though mesmerizing in its beauty, also introduced a whirlwind of emotions. From the profound loss of familiar landscapes to the uncertainty of a drifting future, people grappled with a spectrum of feelings.

Daily Life: Coping Amidst Drifts:

Simple tasks, like fetching groceries or visiting a friend, now came with the anxiety of untethered movement. Many felt a constant sense of vertigo, while others experienced bouts of claustrophobia, feeling trapped in their own floating homes. The unpredictability of the environment, where a sudden gust of wind could shift one's entire world, only added to the stress.

Collective Mourning: The Earth That Was:

Support groups formed, both in physical floating hubs and online platforms, allowing people to share their sense of grief. Old photos, videos, and tales of grounded Earth became cherished treasures. Virtual reality simulations of a world with gravity provided temporary solace to those feeling unmoored

Children's Plight: Navigating a World They Barely Knew:

For children, the shift was even more jarring. Those who had just begun to explore the world found it changed overnight. Schools incorporated emotional well-being lessons, using storytelling and art to help children process their feelings. Play therapy, where kids could enact their fears and hopes, became crucial in providing them a voice.

Solutions in Solitude: The Rise of Meditation and Mindfulness:

The floating world, with its serene vistas and gentle drifts, was conducive to introspection. Many turned to meditation, finding solace in the rhythmic motion of their surroundings. Workshops promoting grounding techniques, even in a groundless world, became popular.

Creativity as Catharsis:

Artists, musicians, and writers channeled the collective emotional upheaval into their creations. Music with the sounds of drifting winds, paintings capturing the ethereal beauty of floating landscapes, and literature exploring the depths of human emotions in this new realm all became therapeutic outlets for both the creators and the audience.

Scientific Intervention: Mental Health in Focus:

Psychologists and neuroscientists began extensive studies on the long-term effects of living without gravity on the human psyche. New therapies were developed, some utilizing virtual reality to simulate grounded experiences, others focusing on group therapy to build communal resilience.

Anchoring Relationships: The Role of Family and Friends:

In a world without physical anchors, relationships became the bedrock of stability. Families devised rituals to stay connected, from daily storytelling sessions to shared meditation hours. Friends formed tighter bonds, often tethering their homes close to create interconnected floating communities.

In the weightless expanse of the floating world, the weight of emotions became even more pronounced. However, the challenges also brought forth the incredible resilience of the human spirit. From innovative therapies to the simple power of shared experiences, humanity navigated its emotional journey, finding pockets of hope, love, and connection amidst the drifts.

Rediscovering Earth: A Journey Down Memory Lane:

The Floating Archives: Preservation of the Past:

As the world took to the skies, there was a collective realization of the importance of remembering Earth as it once was. The Floating Archives were established — vast digital libraries and museums that housed photographs, videos, and artifacts of the grounded world. These repositories became places of pilgrimage, where people could immerse themselves in the sights, sounds, and textures of the Earth they once knew.

Virtual Journeys: Reliving the Grounded Experience:

Technology offered solace in nostalgia. Virtual Reality (VR) systems were developed, allowing users to take strolls down their childhood streets, hike up mountains, or lounge on beaches. These VR experiences became therapeutic, offering an emotional connection to a world that seemed almost mythical.

Story Circles: Oral Histories of the Grounded Days:

Elders, with their rich reservoir of memories from the grounded era, became invaluable storytellers. Communities organized 'Story Circles,' where the older generation would share tales of their past — from mundane memories of walking to school to profound moments of watching sunsets on the horizon. These tales served as bridges, connecting the floating generation to the world below.

Artistic Odes to the Old World:

Artists began to produce works that were reminiscent of the pre-floating era. Paintings capturing serene landscapes, sculptures replicating iconic landmarks, and songs echoing the sounds of the old world became wildly popular. Film directors produced documentaries and feature films set in the grounded world, and these became instant hits, catering to the collective yearning.

The Grounded Parks: Replicating the Old World:

Realizing the emotional importance of tangible experiences, certain large floating landmasses were transformed into 'Grounded Parks'. Here, through a combination of technology and natural elements, environments of the old world were recreated. There were patches of forests, streams, beaches, and even deserts. These parks, though a mere simulacrum, offered people a chance to touch, feel, and relive the sensations of the world that once was.

Educational Expeditions: Teaching the New Generation:

For the newer generation born in the floating world, the grounded Earth was a concept from history books. Schools organized educational trips to the Floating Archives and Grounded Parks. Through hands-on experiences and interactive lessons, children were introduced to the wonders and challenges of the world their ancestors once inhabited.

Restoration Projects: Dreams of Descending:

A few ambitious groups began projects aimed at restoring portions of the Earth to its original state. While these were long-term visions, they garnered significant attention and hope. People contributed to these projects in various ways, be it through funds, research, or labor. The dream of one day descending and reclaiming the Earth became a beacon of hope.

The journey of rediscovering Earth was not merely an exercise in nostalgia but a vital endeavor to understand humanity's roots. As people floated among the clouds, the memories of the ground beneath their feet served as anchors to their identity. The tales, artifacts, and recreated experiences of the old world ensured that while humanity adapted and evolved, they never lost touch with where they came from.

The Return to Normalcy: The Day Earth Regained its Pull:

The Unforeseen Shift:

Dawn broke like any other day in the floating world, but there was something different in the air — a subtle tug, a gentle pull. As the hours progressed, the descent began. Slowly, almost imperceptibly at first, the world started drifting downward. Buildings, plants, animals, and humans felt the embrace of gravity once more, grounding them after what seemed like an eternity in the skies.

The Wave of Realization:

News channels, social media platforms, and community radios buzzed with incredulity and joy. "Gravity's Return!" headlines screamed. Videos of objects gently settling, of children jumping with glee feeling their feet firmly planted, and of water once more flowing downwards went viral.

Immediate Challenges: The Descent Operations:

While the return of gravity was celebrated, it brought immediate challenges. Floating infrastructures had to be stabilized, ensuring they didn't crash or collide. Emergency response teams, prepared from years in the floating world, swung into action, managing the descent and ensuring safety.

Emotional Outpour:

The emotional impact was profound. People knelt, touching the ground, feeling the solid earth beneath their fingers. Parks were filled with families lying on the grass, cherishing the sensation. Tears of joy, relief, and disbelief flowed freely. Impromptu celebrations erupted in streets, with music, dance, and an outpouring of collective emotion.

Scientific Scrutiny: Understanding the Return:

The scientific community was abuzz with activity. Research teams were mobilized to understand the sudden return of Earth's gravitational pull. Early theories ranged from cosmic alignments to core Earth reactions. The world waited with bated breath for explanations.

Economic Revival: Back to Ground Business:

Economies had to readjust. Businesses that had thrived in the floating world found themselves needing to pivot. Traditional modes of transport, agriculture, and infrastructure saw a resurgence. Job markets expanded as reconstruction and groundwork began in earnest.

Environmental Observations: Nature's Response:

Nature responded rapidly. Trees that had adapted to floating began rooting deeper into the earth, rivers resumed their meandering flow, and animals that had mastered aerial navigation now reconnected with their grounded instincts.

Culture and Art: Documenting the Transition:

Artists, writers, and filmmakers took on the task of documenting this monumental transition. Songs celebrating the ground, paintings capturing the moment of descent, and films detailing the return became cultural treasures. Literature explored the duality of human existence in both realms, floating and grounded.

Rebuilding Education: Merging Two Worlds:

Educational curriculums faced an overhaul. Lessons from the floating world were preserved, ensuring future generations would understand the weightless era. But now, they also reintroduced grounded sciences, geographies, and histories, creating a fusion of knowledge from both phases of Earth's existence.

The day Earth regained its pull marked a turning point in human history. The joy of return was complemented by the wisdom gained during the floating era. As people rebuilt, they carried forward lessons of adaptability, resilience, and unity, ensuring a legacy that would be remembered for generations. The return to normalcy wasn't just a return to the familiar; it was a step into a future enriched by the experiences of a world once adrift.